I0493959

Health Risk Concerns of Cell Phones White Paper

By Franklin C. Kostenko, student of MTech, MPEM
12 JAN 2013

Contents

3

The invention of the cellphone was inspired by the trials of war, just as many other useful inventions have been in the past. Although Mobile Telephone Service (MTS) introduced the first cell phone service in 1949, the beginnings of this technology can be traced back to ideas developed 56 years earlier. Italian physicist Guglielmo Marconi succeeded in transmitting wireless signals over a distance of one and a half miles in 1985. Shortly after this remarkable achievement he applied then received the world's first patent for a wireless telegraph system in 1986, created the Wireless Telegraph and Signal Company a year later while perfecting his new technology to be able to send wireless signals for a distance of twelve miles.

In 1901, Marconi pushed the boundaries of science and human knowledge again by proving that the curve of Earth did not affect the successful sending of wireless waves. He demonstrated this by transmitting signals from Cornwall, England to St. Johns, Newfoundland, a distance of over two thousand miles. Over the next few decades Marconi, along with other scientists and researchers, continued to work and do research on wireless transmissions. The first ship-to-shore radio conversation took place in 1922 from the ship S.S. America, a distance of 400 miles. In 1932, the world's first microwave radiotelephone link was established between the summer home of Pope Pious XI in the small Italian town of Caster Gandolfo and Vatican City, a distance of approximately thirty miles. Inter-continental telephone communications between the United States and England occurred in the mid 1930s.

As technology improved, interest in a car phone service developed. The first automobile phone service introduced in 1949 featured big and bulky equipment. The phone required the placement of rather a large receiver, about the size of a piece of luggage, placed in the trunk of the vehicle. In order to be able to hold a phone conversation, the caller had to push a button, and release it to be able to hear the person on the other end of the line. In this way, this phone shared many similarities to two-way radio. The technology that would allow callers to simultaneously talk and listen would not come until much later. What furthered hampered this dawning technology was the necessity of a mobile service company operator routing the call to the desired recipient, as calls could not be direct-dialed.

The introduction of Improved Mobile Telephone Service, or IMTS, in the early 1960s, took the next step in improving service by eliminating the need to employ operators to route calls to phone carriers. In the early 1980s, phones were given full-duplex capability, which allowed callers to simultaneously interact with each other. At the same time, technology had not progressed enough to facilitate smaller and more compact phone designs. Because IMTS technology made phones into what would later be called scanners, lines would often cross each other so that callers would be able to listen in on other conversations. Technology continued to improve to yield the portable, handheld technology that is familiar today, and will most likely continue to evolve in the future. Currently, the more advanced descendant of analog based cellphones, digital technology, has greatly improved the quality of cell phone service, making it much more reliable and secure in addition to allowing carriers to accommodate more callers.

The development of the mobile phone progressed rather quickly from being a device mainly used for communication, to being a smartphone – which is actually a fully functioning computer with a multi-purpose capability. At present, the small size and thermal requirements and limited battery life mean that the computational capabilities of smart phones are still somewhat limited. It is believed that the smart phone in conjunction with cloud computing will extend the capabilities of the current level of technology with the potential of transforming the information technology (IT) and telecommunications industries. Today, technology has advanced to the point where the companies that provide cell phone service are becoming significantly more reliable, for example by dropping fewer calls than they ever have and providing much more extensive coverage over larger physical territories. Despite this remarkable progress, there is still a large portion of the consumer market that is not yet willing to forego the use of land lines (Ruplinger, 2010).

This thesis explored the intersection of the telecommunications and IT technology by investigating an approach known as cloud computing that would increase the speed of mobile phone technology by transferring most of the heavy computing applications to remote computers. In order to test this, a mobile phone application was developed to compare the efficiency of present cloud computing technology to traditional computing technology. Three computational tests were repeated twice - once with cloud computing and once without. Results demonstrated that cloud computing technology is not yet as efficient for these types of tasks as mobile phones.

Origins if the Cell Phone

Martin Cooper - Motorola's visionary pioneer in cell technology - gained his experience with wireless technology as a submarine commander in the United States Navy. The submarine squadrons of World War II provided valuable and convenient testing opportunities for wireless communications. The Germans perfected the technology enabling their submarine fleet to lead all other nations in underwater navigation, range finding systems and communications between submarines. The U.S. Navy paid a price for being unable to keep up with these German advances and felt the life and death urgency to experiment, and develop new technology to catch up. Sonar was one such technological advancement that was invented through these circumstances. Sonar utilizes underwater sound waves that are sent out and returned to their origin after striking off solid surfaces. The speed with which they returned signals determines the distance and depth of the given object from the submarine. RADAR would come a generation later, relying on the same basic concept out of the water that sonar had developed under water.

Radio Detection and Ranging (RADAR) changed the world by making weather forecasting far more accurate, thus enabling the military to be more efficient and facilitating safer and faster air and sea travel. RADAR was discovered by physicist Heinrich Hertz who began experimenting with radio waves in his German laboratory in 1987. Hertz found that although radio waves could be transmitted through some material, they would go though others and would bounce back instead. The RADAR race took off by the 1920s as Great Britain, Germany and the United States all competed to be the first to develop the new technology (2). In the United States, Cl. William R. Blair, US Army director of the Signal Corps Engineering Laboratories, was the first American to receive a patent for RADAR (4).

The invention of RADAR could have saved the Titanic had it been around in 1912, yet ironically the large public outcry that followed the sinking of the "unsinkable" Titanic more than anything else may have spurred the discovery, development and perfection of RADAR technology pushing its transition from military into robust civilian use (1). It took only twenty years before scientists around the world began to discover the practical use of radio waves to detect and locate objects (3) such as icebergs, storms and other ships at sea.

American sailors serving on the first ships to use radars on deck noticed a warm sensation when standing in front of the radar antennas for even a few minutes. In the cold, damp, and windy weather conditions out in the open sea, the warmth generated by radar was considered to be a positive aspect of RADAR. This fact did not elude scientists, who utilized the heat generating characteristic of RADAR to develop the first microwave ovens, originally called radar ranges, which heated food by sending wavelengths of 915 MHz into the oven, similar to many of today's cellular phones.

In the beginning, some cell phones relied on wavelengths similar to those used by earlier microwave ovens and by primitive radar. The only noticeable difference was the power (wattage) that was needed to operate radar on a microwave oven. Today's cell phones send pulsed peak signals that operate on average at about one watt (5). The current 3G and 4G smart phones that are on the market utilize the same radar waves as microwave ovens.

RADAR technology continued to be actively used by the military, and during World War II gave the Allies a decisive advantage over the Japanese, who, at the time, did not have access to operational radar. Americans had the element of surprise on their side because of this advantage. Having advance notice of an air raid often meant the difference between victory and defeat for the Allies (1). RADAR for ship navigation and airplane navigation through stormy weather did not become available commercially until after World War II. In the decade after the war, radar became increasingly used at airports to ensure safe approach and landing.

In 1946, the military achieved a breakthrough in the use of RADAR, when soldiers at US Army Camp Evans conducted an experiment aiming a concentrated high frequency "Diana" radar beam at the moon. The beam took only two and a half seconds to produce an audible ping over their receiver loudspeaker traveling at the speed of light (186,000 miles per second). This achievement ushered in the development of satellite communications and missile guidance systems that are now commonly in use (1). Near the end of World War II, as the toll on England's air force mounted, coming up with a platform level warning capability became an urgent need. In order to fit radar and it's antennas onto planes, scientists were forced to make them smaller and more portable.

Since World War II, the American military has used advanced radars in every war, helping to protect soldiers, sailors and civilians in the battle space, making air and highway travel safer and faster, as well as, making accurate weather, which has greatly contributed to making travel safer.

The Technology

Wireless phones receive their signals from elevated manmade towers. A cell is defined as the area (several miles) around a tower in which a signal can be picked up and received from the tower. Each cell has a base station that consists of a tower and a small building containing radio equipment. A cell phone is a full-duplex device, which means that they utilize one frequency for talking functions and a second, separate frequency for listening capabilities. The division of a city or region into a series of limited cells allows extensive frequency reuse across a city, so that millions of people can utilize cell phones simultaneously. Cell phones operate within cells, and have the ability to switch between cells as users travel and move around geographically. Cells give cell phones incredible range, as users are able to drive hundreds of miles, while maintaining conversations the entire time due to a vast cellular network.

Cell phones provide an incredible array of functions. Depending on the cell-phone model, one can:
- Store contact information
- Create task or to-do lists
- Schedule appointments and set reminders
- Utilize the built-in calculator for simple math

- Send or receive e-mail
- Obtain information (news, entertainment, stock quotes) from the internet
- Play games/listen to music
- Watch TV/read books
- Send text messages/Skype
- Integrate other devices such as PDAs, MP3 players and GPS receivers

A single cell in an analog cell-phone system uses one-seventh of the available duplex voice channels and unique frequencies in order to avoid collisions. Any given cell phone carrier typically is able to receive approximately 832 available frequencies in a city, with every cell phone using two frequencies per call, or what is known as a duplex channel. This means that each carrier has 395 voice channels and 42 control channels allowing 60 callers to simultaneously talk on their cell phones without straining the network.

Analog cellular systems are considered first-generation mobile technology, or 1G. With digital transmission methods (2G), the number of available channels increases. For example, a TDMA-based digital system can carry triple the amount of calls of an analog system, so that each cell has about 200 channels available.

Cell phones employ low-power transmitters. Many cell phones have two signal strengths, 0.6 watts and 3 watts. In addition to this, the base station also transmits signals at low power. Low-power transmitters have two advantages. The first advantage is that transmitted signals are confined mainly within the cell, allowing two callers to reuse the same frequencies across the same city. The second advantage is that the power consumption of the cell phone, which is normally battery-operated, is relatively low. Low power eliminates the need for large batteries, therefore making compact, handheld cellular phones a possibility.

The cellular approach necessitates a large number of base stations in any given region, many of which have hundreds of towers. The vast number of cell phone users in a region enable per user costs to remain low. Each carrier runs a central office in each region called the Mobile Telephone Switching Office (MTSO), that handles all phone connections to the land-based phone system, and controls all of the base stations.

Cell phones have special codes associated with them that identify the phone device, the phone's owner and the service provider. The various cell phone codes in existence are as follows:

1. Electronic Serial Number (ESN) is a unique 32-digit number programmed into the phone when it is manufactured.

2. Mobile Identification Number (MIN) is a 10-digit number derived from the device's phone number.
3. System Identification Code (SID) is a unique 5-digit number assigned to each carrier by the Federal Communications Commission (FCC).

ESN is a permanent part of the device, while both MIN and SID codes are programmed into the phone at the time of purchase.

Cell phone carriers use Mobile Telephone Offices (MTSOs) as bases that station cell sites connect to. This is actually a sophisticated computer that monitors all cellular calls, keeps track of the location of all cellular-equipped vehicles traveling within the system, arranges hand-offs between towers, and keeps track of billing information. The MTSO, in turn, interfaces with the Public Switched Telephone Network (PSTN) through a connection to a Control Office. Carriers operate within the PSTN - a network of the world's public circuit-switched telephone networks - in much the same way that the Internet is the network of the world's public IP-based packet-switched networks. Originally a network of fixed-line analogue telephone systems, the PSTN is now almost entirely digital, and includes mobile as well as land lines.

When first powered on, cell phones listen for an SID on the control channel, which is a special frequency that the phone and base station uses to communicate call set-up and channel changing. If the device cannot find any control channels to listen to, it recognizes that it is out of range and displays a "no service" message. Once a SID is received, the device then compares it to its programmed SID. If the SIDs match the device recognizes that it is within its home network. Along with the SID, the phone also transmits a registration request, while the MTSO

14

tracks the device's location in a database, so that it can properly route incoming calls to the intended devices.

Once the MTSO receives a call, it attempts to locate the intended device within the database to find the appropriate cell. It then chooses a frequency pair that the device will utilize in that cell to support the call. Next, the MTSO communicates with the device through the control channel to communicate which frequencies to use, and upon device and the tower switching to the supported frequencies, the MTSO connects the call.

As devices move towards the edge of their cells, the base station notes the diminishing strength signal. Meanwhile, the base station within the adjacent cells registers the increasing signal strength. The two base stations coordinate with each other through the MTSO, and at some point, the device receives instructions on the control channel to change frequencies to the new cell.

If a user moves from one cell to another, and another service provider covers this new cell, the device is handed off to that service provider. If the SID on the control channel does not match the device's programmed SID, the device recognizes that it is roaming. The MTSO of the roaming cell contacts the MTSO of the home network, which runs the SID through its database in order ensure the validity of the SID. The home network verifies the device as belonging to the local MTSO where its movements are further tracked. The entire process takes only a few seconds to complete. Some devices actually display the word "roam" on their screens to inform users and to help them prevent roaming charges. In the case of users traveling internationally, they need to obtain a phone that is compatible with both their home networks, as well as those abroad as different countries use different cellular access technologies.

3G technology is intended for use by the true multimedia cell phones, typically called smart phones, which feature increased bandwidth and high speed transfer rates to accommodate Web-based applications and phone-based audio and video files. 3G comprises several cellular access technologies as follows:

1. CDMA2000, which is based on 2-G Code Division Multiple Access.

2. Wideband Code Division Multiple Access-UMTS (WCDMA-UMTS), which allows users to simultaneously transmit data at different rates time packets in W-CDMA interface. UMTS networks support all current second-generation services and numerous new applications and services.

3. Time Division Synchronous Code Division Multiple Access (TD-SCDMA), which TD uses the Time Division Duplex (TDD) mode, that transmits uplink traffic (traffic from the mobile terminal to the base station) and downlink traffic (traffic from the base station to the terminal) in the same frame but in different time slots. This means that the uplink and downlink spectrum is assigned flexibly, dependent on the type of information being transmitted. When asymmetrical data like e-mail and internet are transmitted from the base station, additional time slots are used for downlink than for uplink. A symmetrical split in the uplink and downlink occurs with symmetrical services like telephony.

Digital cell phones utilize the same radio technology as analog phones, but they use it in a different way. Analog systems do not fully process the signal between the phone and the cellular network as they cannot be compressed and manipulated as easily as a true digital signal. Digital phones convert users' voices into binary information of 1s and 0s and then proceed compress them. This compression allows anywhere between three and 10 digital cell-phone calls to occupy the space of a single analog call. Many digital cellular systems rely on frequency-shift keying (FSK) to send data back and forth over AMPS. FSK uses two frequencies, one for 1s and the other for 0s, alternating rapidly between the two to send digital information between the cell tower and the phone. Clever modulation and encoding schemes are required to convert the analog information to a digital signal, compress it and convert it back again while maintaining an acceptable level of voice quality. Digital cell phones need to contain a lot of processing power

Cell Phone Components

There are a few individual parts inside every basic digital cell phone including:
- a circuit board containing the "brains" of the phone,
- an antenna,
- a display screen,
- a keyboard,
- a microphone,
- a speaker, and
- a battery.

The circuit board is comprised of several chips and controls the system. The analog-to-digital and digital-to-analog conversion chips translate the outgoing audio signal from analog to digital and the incoming signal from digital back to analog. The digital signal processor (DSP) is a highly customized processor designed to perform signal-manipulation calculations at high speed. The microprocessor handles all the functions of the keyboard and display, deals with command and control signaling with the base station and also coordinates the rest of the functions on the board. The Read Only Memory (ROM) and Flash Memory chips provide storage for the phone's operating system and customizable features, such as the phone directory. The Radio Frequency (RF) and power section handles power management and recharging, and also processes the hundreds of available FM channels. Finally, the RF amplifiers handle signals traveling to and from the antenna.

Over the last few years, the display has become larger in size to accommodate the increasing number of features and applications being performed by cell phones. Most current phones offer built-in phone directories, calculators and games. In addition to this, the majority of phones incorporate some form of personal digital assistant (PDA) or Web browser. Some phones store certain information, such as the SID and MIN codes, and internal Flash memory, while others use external cards. Finally, cell phones incorporate minute speakers and microphones.

A cell-phone tower is typically a steel pole or lattice structure that rises hundreds of feet into the air. The box houses the radio transmitters and receivers the signals that allow the tower communicate with the phones within its cell. Transmitters and receivers then connect with the tower's antennae through thick cables. The tower along with its cables and equipment are all heavily grounded.

Transmitters encode the sound of users' voices onto continuous sine waves, which are simply a type of continuously varying waves radiating out from the antenna and fluctuating evenly through space. Sine waves are measured in terms of frequency. Once the encoded sound has been placed on the sine wave, the transmitter sends the signal to the antenna, which then sends the signal out.

Cell phones use built-in low-power transmitters and operate on about 0.75 to 1.0 watts of power. The position of a transmitter inside a phone varies depending on them manufacturer, but it is typically placed in close proximity to the phone's antenna. The radio waves sending out the encoded signal comprise of electromagnetic radiation propagated by the antenna. The antenna's function in any radio transmitter is to launch the radio waves into space, and in the case of cell phones, the tower's receiver picks up these waves.

Broadband Signal Distribution

The term broadband refers to the wide bandwidth characteristics of a transmission medium and its ability to transport multiple signals and traffic types simultaneously. The medium can be coax, optical fiber, twisted pair or wireless (used in hand-held devices). In contrast, baseband describes a communication system in which information is transported across a single channel. Prior to the invention of home broadband, dial-up Internet access was the only means by which users were able to access the Internet and download files such as songs, movies, e-mails, etc. It typically took anywhere from 15–30 minutes to download one song (3.5 MB) and over 28 hours to download a movie (700 MB). Dial-up Internet was also considered very inconvenient as it would impair the use of the home telephone line forcing users to consider the purchase of additional phone lines.

In 1997, the cable modem was introduced, although the common use of broadband did not experience a significant rise in use until 2001. Having a broadband connection enabled users to download and send files significantly faster when using a dial-up. As with many new technologies, most consumers were unable to afford the cost of faster Internet service. Because high costs were no longer a factor, by 2004 most average American households considered home broadband service to be affordable. Since its inception, broadband has continually strengthened and enabled connection speeds continue to rise.

Different criteria for the term "broadband" have been applied in different contexts. The origin of the term lies in physics, acoustics and radio systems engineering, where it had been used to signify wideband. However, the term "broadband" became popularized in the 1990s, as a vague, but popular and highly effective, marketing term for Internet access, even as accessed on today's smart cell phones.

Space Based Communications

According to United States National Space Policy, satellites contribute to increased transparency and stability among nations and provide vital communication paths for avoiding potential political conflicts. The use of space based systems helps save lives by employing warnings of impending natural disasters, enables agriculture and natural resource management to be more efficient and sustainable, and provide global access to advanced medicine, weather forecasting, geospatial information, broadband and other communications.

It is critical that space based positioning, navigating, and timing systems are maintained and continually enhanced. Modern smart phones rely heavily on the use of global navigation satellite systems (GNSS). Other peaceful civil uses include Global Positioning systems (GPS) and its government provided augmentations, which is provided free of direct user charges. The smart phones that are currently in use have the ability to engage foreign GNSS providers to promote transparency in locating and timing requirements, a function also called foreign positioning, navigation, and timing (PNT) services. The telecommunications industry is actively pushing to increase these public-private partnerships. The advent of smart phone technology has brought this space based situational awareness into the palm of users' hands. The current space systems have the ability to conduct radio-frequency surveys from the Earth's orbit, however none of the collected data is currently available for reproduction in an unclassified environment. Space systems today allow individuals around the world to see one another with clarity, communicate with certainty, navigate with accuracy, and operate with assurance (National Security Space Strategy 2011). In conclusion, space is becoming increasingly more congested, which includes the radio-frequency spectrum, and well as more contested and basically competitive.

Cell phone devices are wholly dependent on these space-based systems and the localized terrestrial supporting infrastructure for smart phone operations. Together, they create the globally connected cellular network domain. Cellular phones and satelines are subject to environmental considerations, mainly space weather due to the radio-frequency disturbances that may be encountered - interference affected by solar flares, charged particles, cosmic rays, and the Van Allen radiation belts. These and other natural phenomena affect communications, navigational accuracy, performance on unit sensors, and even may cause electronic failure. Electromagnetic interference (EMI) may be the result of such natural occurring phenomenon. No geographic boundaries exist today for what is commonly called the Global Information Grid (GIG).

Possible Health Side Effects

One of the most common concerns that cell phone users have is the possible health risk that could occur as a result of cell phone use. Cell phones are typically placed against user's head and ear in a telephone conversation. This position makes it possible for a portion of the radiation to be absorbed by human tissue because all cell phones emit some electromagnetic radiation. Given the close proximity of the phone to the head, it is possible for the radiation to cause a certain degree of harm to the user. What is being debated in scientific and political realms is just how much radiation is considered unsafe, and the existence of any potential long-term effects of cell-phone radiation exposure.

There are two types of electromagnetic radiation: ionizing radiation and non-ionizing radiation. Ionizing radiation, found in gamma rays and X-rays, contains enough electromagnetic energy to strip atoms and molecules from the tissue and alter chemical reactions in the body. Humans typically protect their bodies with lead vests when they have to be exposed to this type of radiation. In contrast to this the non-ionizing radiation found in visible light, microwave radiation, and radio-frequency energy is considered to be much safer. This form of radiation does emit some heat, but typically not enough to inflict long-term damage to human tissue.

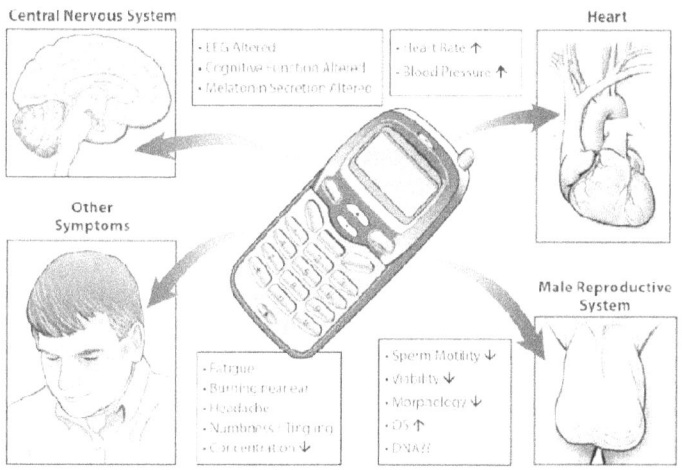

On its Web site, the FDA asserts, "the available scientific evidence does not demonstrate any adverse health effects associated with the use of mobile phones" (FDA, 2013). However, this does not necessarily mean that the potential for harm does not exist. Radiation can still cause damage to human tissue if there is exposure to high levels of RF radiation, according to the FCC. RF radiation has the ability to heat human tissue in much the same way as microwave ovens heat food. Damage to tissue can be caused by exposure to RF radiation because the human body is not equipped to dissipate excessive amounts of heat. The eyes are particularly vulnerable due to the lack of blood flow. The added concern with non-ionizing radiation - the type of radiation associated with cell phones - is that it has the potential to cause long-term effects. Although it may not immediately cause damage to tissue, scientists are still unsure about whether prolonged exposure could create health problems. This continues to be a polarizing issue today as the number of cell phone users continues to increase. Some of the diseases that are suspected of being potentially linked to cell-phone radiation are cancer, brain tumors, Alzheimer's, Parkinson's, fatigue, and headaches. In addition, cell phones may cause use addictions.

The risks as understood today: genetic damage

There is growing evidence that the negative effects of microwave radiation accumulate over time. In order to work, cell phones have to send and receive signals from a base station, connecting with all other cell phones in an area to form a web of information carrying radio waves. The Federal Telecommunications Act of 1996 essentially prevents local authorities from considering health concerns in deciding where towers are to be placed. The FCC and the FDA, which regulate cell phone carriers, claim that because the cell phone produces no heat, it is safe for all to use. The Bioelectromagnetics Society Newsletter has published over 40 separate reports of the risk and factual cell swelling and suppressions as well as

'DNA double strand breaks' from the use of smart phones. They report that the use of cell phones can cause skin cancer, especially at the favored ear for cell communication.

The United States Environmental Protection Agency has passed the subject of radiation on to the FCC, however they do note there is a large volume of cell phones retired each year, likely up to 150 million per year. The FCC has never employed medical expert as part of its staff. The reality is that, in their circuitry, batteries, and liquid crystal displays, cell phones can contain toxics like arsenic, nickel, coltan, beryllium, cadmium, copper, and lead. Their plastic casings have also been treated with brominated flame-retardants. (EPA 2013) All these facts are true, yet they do not mention the cell phones principal health risk. The EPA has entire web site containing information on disposal techniques and upgrading recommendations, however, they currently have left out the potentially adverse health effects to humans as a result of cell phone use. Currently, the FDA tests microwave ovens for safety, but does not do anything in terms of testing cell phones.

Although most users disregard possible side effects of cell phone use, cell phone manufacturers are beginning to take notice of and respond to the threat of possible risks. As of 2010, the Motorola V195 model included a warning to keep the cell phone one inch from the user's body; the BlackBerry 8300, 0.98 of an inch; the Nokia 1100, one fourth of an inch; and the iPhone five-eighths of an inch. The new Verizon Droid Eris cell phone contains a "Product Safety and Warranty Information" booklet. On page 11 of the booklet, it advises users "that no part of the human body [is to] be allowed to come too close to the antenna during operation of the equipment". A customer query about this was referred to an online appendix, which explained on page 219, "To comply with RF exposure requirements, a minimum separation distance of 1.5 cm must be maintained between the user's body and the handset, including the antenna." Similarly, on page 5 of the iPhone 4 user booklet there featured a warning about exposure to radio frequency energy:

> "For optimal mobile device performance and to be sure that human exposure to RF energy does not exceed the FCC, IC, and European Union guidelines always follow these instructions and precautions: When on a call using the built-in audio receiver in iPhone, hold iPhone with the dock connector pointed down toward your shoulder to increase separation from the antenna. When using iPhone near your body for voice calls or for wireless data transmission over a cellular network, keep iPhone at least 15 mm (5/8 inch) away from the body, and only use carrying cases, belt clips, or holders that do not have metal parts and that maintain at least 15 mm (5/8") separation between iPhone and the body" (Apple, 2013).

When rats are exposed to radio-frequency-radiation, the DNA in their brain cells show broken helical segments. These altered brain cells are the same as those known to occur in cancer. To remain healthy, DNA must remain intact. Long term effect on children are still unknown in the United States, however, Russia has conducted some advanced testing which yields potential clues. For 5 years, in a relatively short study, the Russian

Government examined two groups of children, a control group which did not regularly used cell phones, and a test group that did. Those in the test group experienced dramatic differences from those in the control group after only five years. The subjects in the test group now have a host of what may be called functional problems, difficulties with learning and behavioral issues.

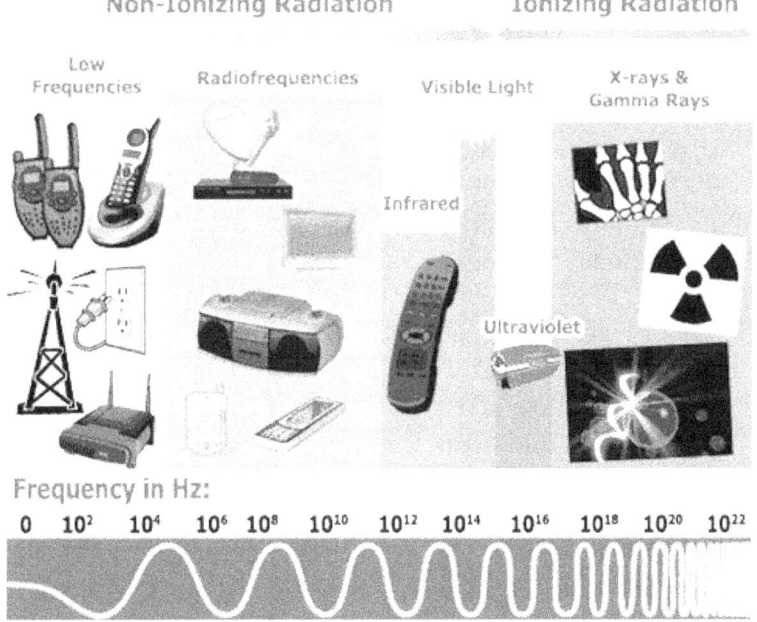

Description Hazard Probability

The phrase "heart stopping phone calls" was coined in the early 1990's. Increasingly, reports of users' pace makers being stopped by cell phones surfaced. The FDA kept track of such reports and noted that heart defibrillators were also reported to be failing after cell phones were used in close proximity. None of the cases featured reports of cancer, and all involved electronic interference. In Italy, this prompted researchers to test digital cellular phones in a laboratory and in patients with pacemakers. When the phones operated in close proximity to the pacemakers, about 4 inches, interference of some type was found in half the instances (Carlo, et al. 2001).

Over the last decade, recent experimental studies showed that the reduction of insect reproductive capacity, when it close proximity to cell phone radiation was due to DNA damage, actin cytoskeleton damage, and cell death induction in the reproductive cells (Gonads). GSM 900 and 1800 MHz mobile telephony radiation has been found to reduce insect reproduction by up to 60%. With only five mines of typical exposure a day both males and females were affected. GSM 900 MHz radiation was deemed to be more bioactive than GSM 1800 MHz under actual conditions mainly due to the fact that the GSM 900 is emitted at double the output power than GSM 1800 (GSM 1900).

Electromagnetic stress seems to be more bioactive than other previously tested stress factors like poor nutrition, heat, or chemical stress, including DNA damage to a higher degree on insect reproductive cells. The effect at the cellular level seems to be due to irregular gating of ion channels on the cell membranes caused by the EMFs, leading to disruption of cell's electrochemical balance and function. Highly reported observations during the last years regarding reduction of bird and insect populations, especially bees, can be explained by decreases in their reproductive capacity as described in our experiments. Symptoms referred to as "microwave syndrome" (headaches, sleep disturbances, fatigue etc.), among people residing around base station antennas, can possibly be explained by cellular stress induction on brain cell or even cell death induction on a number of brain cells (Panagopoulos, 2011).

Other Hazards

With the rapid proliferation of social networking websites such as Facebook, Linked In, Twitter, MySpace and Bebo humans are eager to stay connected with peers and are open to sharing personal information with others (6). This led researchers to OLS or Opportunistic Localization System for smart phones & devices. OSL have the ability bridge the gap between outdoor and indoor localization via Wi-Fi, GPS and other networked systems. This allows mobile devices to locate others like them almost anywhere in the world. Currently manufactured mobile devices also have inertial sensors build-in (mostly accelerometers and or compasses) that can be used as extra localization information using the pedestrian dead reckoning (PDR) principle, which is used to estimate distance.

Users can now perform activities that traditionally grounded them to their homes and offices while being mobile. While traveling, one can email, video conference and watch streaming videos using your smartphone. Currently, smart phones and other devices offer transfer speeds of about 3.6 Mbits per second and even more, which can make data transfer seamless and downloads relatively fast.

Individuals use cell phones everywhere, even when doing so poses a danger, such as driving. The passages of laws against using a smart phone cell while driving seem to make little difference. Some studies have shown, driving while on the phone or texting is more dangerous then driving under the influence. Constant cell phone use may be creating behavioral changes in the user and influencing young adults in a particularity negative way. Using a phone while driving increased the risk of an accident by eight times.

In contrast to the United States, other nations are aggressively restricting cell phone use issuing warnings about their potential hazards. There are superhot effects or super thermal effects that occur in very tiny spaces that are exposed to radio frequency radiation hotspots. Researchers feel that the medical world is already seeing damage because heat is occurring in the tiniest spaces that cannot be picked up by our current systems of measurement (Davis, 2010).

Risk Management Options for Health and Hazard Risk Management: Conclusions and Recommendations

It seems clear that the use of cell phones is now commonly considered voluntary. A voluntary risk is much more acceptable to users than an imposed risk. Risks that individuals can take steps to control are more acceptable than those they feel are beyond their control. These outrage factors are not distortions in the public's perception of risk. They help to explain as to why the public fears pollutants in the air and water more than they do geological radon or microradio waves. The problem is that many risk experts resist the use of the public's "irrational fear "in their risk management. A problem exists in the perception of risk because the experts and lay views differ. The experts usually base their assessment on mortality rates, while the lay fears are based on the aforementioned "outrage" factors. One additional example is the ongoing concern for the risks involved with cigarette smoke. Another effort must be made to decrease the public's concern with low to modest hazards, i.e., risk managers must diminish "outrage" in these areas. In addition, individuals must be treated fairly and honestly so that trust is built between exposed communities and the risk managers and responsible parties (Theodore, 2012).

If individuals choose to use mobile phones, they can minimize your exposure to radio waves by:

- Keeping cell phone calls short and to a minimum,
- Considering relative SAR values when purchasing a new phone,
- Using speaker setting and keeping the handset away from the head and body,
- Using a low power wire-less headphone with a low power Bluetooth emitter.,
- Using a wired headset,
- Considering the use of airplane mode,
- Placing the phone away from the body as much as possible,
- Keeping the unit powered off, when keeping it next to the waist,
- Only using the unit when signal strength is good,
- Avoiding using the devise in a box environment, such as an elevator,
- Texting more often,

- Advising pregnant women to keep cell phones away from their abdomens,
- Protecting babies from cell phone exposure,
- Keeping the cell phone off when in your pocket,
- Using a landline at home and not a cordless phone,
- Avoiding texting or speaking on a cell phone when driving,
- Keeping all Bluetooth and Wi-Fi settings off when not in use, and
- Reading all user manuals and the FCC web site for updates on health risks from cell phones at www.fcc.gov/cgb/cellular.html

Cost/Benefit Analysis

While it cannot be denied that cell phones and the associated technology has propelled every single industry in the world and advanced the history of humankind, there are considerable risks that need to be addressed. Every year, insurance companies pay out thousands of dollars to cover accidents that resulted due to the careless use of cell phones and other hand held technology. Beyond this, the much more serious issue of loss of life needs to also be considered. Research has proven that using a cell phone during driving results in much slower reaction times than driving under the influence. When drivers kill another person due to texting or cell phone use, this has the same effects as when something similar results from the combination of driving and alcohol and drug use. These events cannot be reversed, but even though they are serious, not enough drivers are acknowledging the risks. Law enforcement officials need to work with drivers in order to pass laws that will save and protect lives. The truth is that society has become so accustomed to its dealings being constantly on the go and instantly accessible that humans are willing to risk their lives and those of others in an attempt to do what they seemingly feel obligated to.

Changes can be implemented not only with laws, but also cell phone designs. Cell phones that operate much easier when they are in speaker mode would facilitate users being more willing to use this function. This would also help alleviate any concerns regarding adverse health effects as a result of cell phone use. Technology has come a long way since the beginnings of the 20th century, and it can certainly go farther, in a bid to improve itself, be more functional, and help protect human life.

References:

1) Carlo, G., and Schram M., (2001). "Cell Phones, Invisible Hazards in the wireless Age", Carrol & Graf, New York, NY 84.

2) Panagopoulos, D.J. (2011). Edited by Barnes, M.C., and Meyers N.P. "Mobile Phones, Technology, Networks and User Issues, Media and Communications", Nova Publications, New York, NY 34, 35

3) Theodore, L and (2012). Environmental Health and Hazard Risk Assessment: Principles and Calculations (Page 311). CRC Press. Kindle Edition.

4) The Technology that changes the future by Wendy Rejan, US Central Command Historian Dec 2007

5) The June 2007, Monmouth Message Magazine, Article on Radar Development

6) The New York Times, http://www.nytimes.com/learning/general/onthisday/big/0415.html

7) Electronic Design, Vol 24, number 4, Feb 16,1976 of Hayden Publishing Comp Inc. NJ

8) Davis, D. (2010) *Disconnected*, Penguin Group, New York, USA p40.

9) Barnes, M.C. and Meyers, N. P. (2012) Mobile Phones, Technology, Networks and User Issues, Nova Science Publishers, New York p192

www.ingramcontent.com/pod-product-compliance
Lightning Source LLC
Chambersburg PA
CBHW070728180526
45167CB00004B/1668